Jane foster's
I LOVE Green

templar books

I love green!

Green always wins.

Everything I have is green

and that's what makes me grin!

It's the colour of my dinosaurs when I line up my toys.

A roaring stegosaurus, making lots of noise!

ROAR!

I love riding on my bike –
of course my bike is green . . .

. . . I also have a shiny bell,
the brightest green you've seen!

I have a little gecko,
(other pets are SO much duller).

We have one thing in common – green's our favourite colour!

Avocados,

apples,

a bunch of **grapes** for me!

This colour tastes the best!

(Apart from Dad's green tea . . .).

I always like to count the peas sitting on my plate:

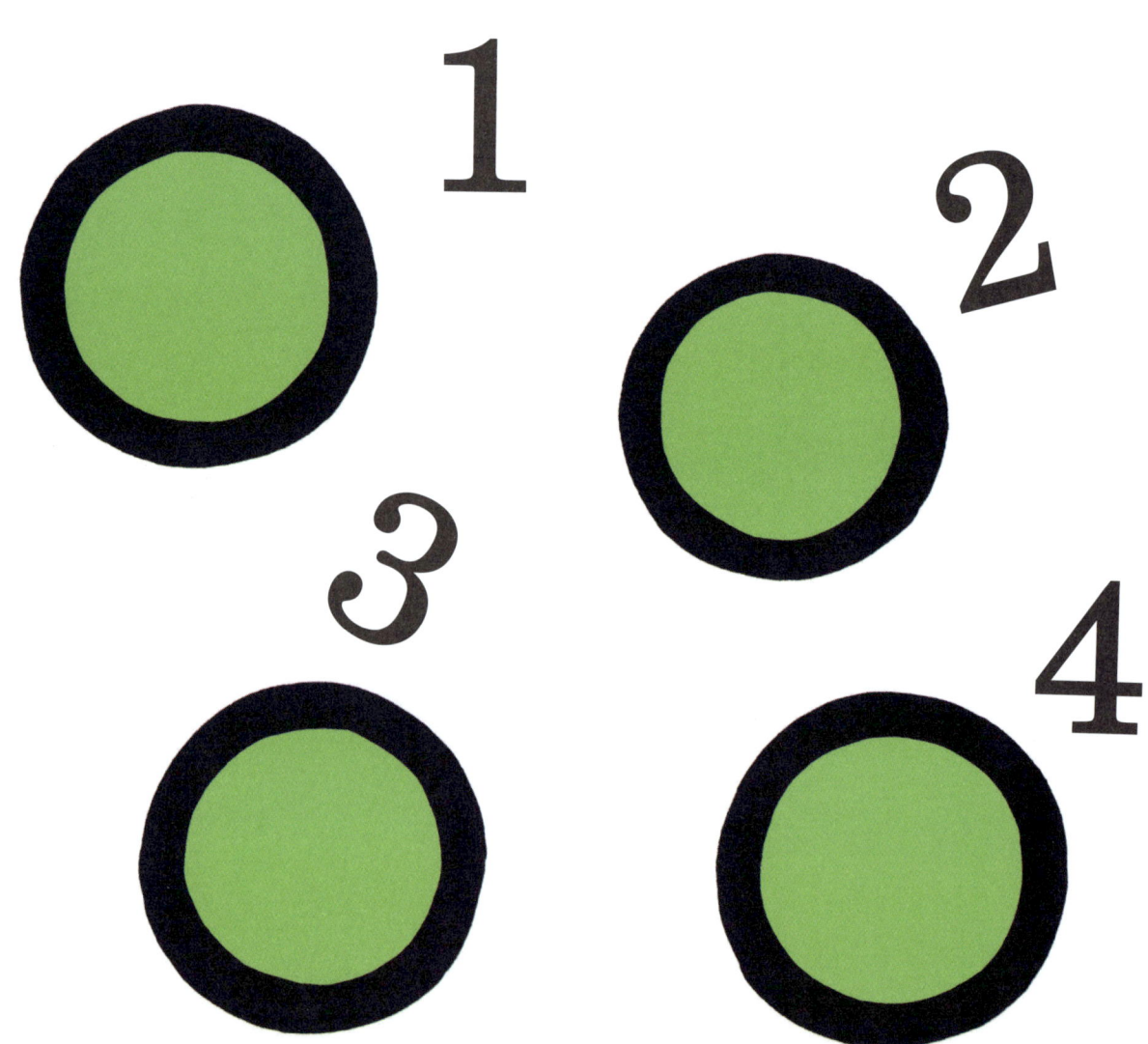

My mummy has a name for me
(it really isn't mean).
She says it when she cuddles me . . .
She calls me her

'green bean'!

I love green!

It's better than the rest.

Now everyone join in and say,
green is the best!

A TEMPLAR BOOK

First published in the UK in 2024 by Templar Books,
an imprint of Bonnier Books UK
4th Floor, Victoria House,
Bloomsbury Square, London, WC1B 4DA
Owned by Bonnier Books
Sveavägen 56, Stockholm, Sweden
www.bonnierbooks.co.uk

Text and illustration copyright © 2024 by Jane Foster
Design copyright © 2024 by Templar Books

1 3 5 7 9 10 8 6 4 2

All rights reserved

ISBN 978-1-80078-695-0

Edited by Ruth Symons
Designed by Genevieve Webster
Production by Ché Creasey

Printed in China